Mindful Triage

Positive Psychology and Life in the ER

Mindful Triage

Positive Psychology and Life in the ER

Joshua Harvey **RN, CEN, CCRN**

ISBN 978-1-257-15694-8

Cover Design by Chris Scheidies
scheidies@gmail.com

Expressive Aphasia Publishing Company
crazypsychnurse@gmail.com

Acknowledgements

Thanks to you, my dear reader, for tolerating the writing style that betrays my multiple-personality disorder. This comes from years of routine teaching Mental-Health Nursing. I drive my four kids to school, and then sit in my office staring blankly at the wall while my coffee brews. I might spend the day lecturing about schizophrenia or teaching Advanced Life Support for Experienced Providers. Whatever, routine. Then the weekend comes and with it comes the enjoyable cacophony that is the Emergency Room. I have never said that "nothing surprises me". Shit astounds me every single shift I work. That's why we love the ER.

Special thanks go out to all my nursing students who I have bounced some of these stories and ideas off of. I am indebted to Larry Harvey, Mark Harvey, Peg Radke, Pat Muth, and Jeremy Nickel for reading the very first draft and offering encouragement and insight. Thanks to Chris Scheidies for the great cover. Thanks to my wife, Amy for hanging out with me these last fifteen years and tolerating my writing bipolarity. Thanks to my four kids Naomi, Faith, Isaac, and Nathan for making life rich and fulfilling.

Contents

1   Open to Life versus Self-Absorption and Somatization

These days it seems like any idiot with a laptop computer can churn out a business book and make a few bucks. That's certainly what I'm hoping...

Scott Adams

... and that's when I decided that I wanted to be a writer, because there were no requirements. All you had to say was, "I am a writer," and you became one. You didn't even have to write anything. You could just sit in a coffee shop with a notebook and stare into space, with a slightly bemused look on your face, judging the weight of the world with a jaundiced eye. As you can see, you can be completely full of shit and still be a writer. Okay, maybe that's the one requirement.

Lewis Black

I am no fan of books and chances are, if you're reading this, you and I share a healthy skepticism about the printed word. Well, I want you to know that this is the first book I've ever written, and I hope it's the first book you've ever read. Don't make a habit of it.

Stephen Colbert

I have two quick exercises for you to do. WAIT! Don't put the book back on the shelf! You don't have to do anything if you don't want to. Consider doing these just for fun. Throughout this book, I will have a few fun little exercises to do which will (I hope) help us with engagement. It seems that we tend to get more out of things that we put a little more into. So, exercise number one: Write a quick paragraph about what you saw on your way in here today! If you are thumbing through this at a college bookstore, because you are in the nursing section and its attractive cover caught your eye: what do you remember about the parking lot? Was there an asshole in a sleeveless wife-beater T-shirt standing outside a 40-year-old station wagon with a NASCAR bumper-sticker, smoking and yelling at his kids? Write that down. What was your impression? (Don't be too harsh, it might have been this author. Haha, just kidding. I quit smoking a long time ago). Anyway, if you are on Amazon.com sitting in your basement looking at this book online, write down what you remember just prior to walking into this room.

Another exercise that I would like you to try is to write down in one sentence, a 'definition' of yourself. Who are you as a human being?

Strange question? It kind of is... The point of both of these exercise are of some psychological importance. Many therapists (especially those of a more Eastern school of thought) will grade your level of mental health/illness based on how 'open' to experience versus how 'closed' to life you are. Contracted, is another word for that inward-looking, self-absorbed state that we see in high anxiety. A therapist might ask how many cars were in the parking lot on the way in to the building. This allows a glimpse into the level of self-absorption of an individual. Somatization, the heightened awareness of our physical aches and pains, is an example of being self-absorbed.

Have you ever had a severe back-ache as you were tiredly driving home through the snowy mountains? You try to shift your weight in the seat and the pain persists. You are trying to distract yourself from the pain by talking on your phone but the back pain is your only real focus. Suddenly you have a near-death experience as a rock slide hurls elephant-sized boulders into the road ahead. Instantly, you're careening off the icy, winding mountain-road, sliding sideways and stopping just before skidding off of a 1,000 foot drop into the raging, white-capped river below. Ever had that happen?

Isn't funny how you had forgotten about your back pain for just a moment. Suddenly the bitching and moaning that you were doing just prior does not seem to be quite as relevant. You are no longer in a contracted, self-absorbed state.

For the last half century psychology has been consumed with a single topic only—mental illness—and has done fairly well with it. Psychologists can now measure such once-fuzzy concepts as depression, schizophrenia, and alcoholism with considerable precision. We now know a good deal about how these troubles develop across the life span, and about their genetics, their biochemistry, and their psychological causes. Best of all, we have learned how to relieve these disorders. By my last count, fourteen of the several dozen major mental illnesses could be effectively treated (and two of them cured) with medication and specific forms of psychotherapy. .... the time has finally arrived for a science that seeks to understand positive emotion, build strength and virtue, and provide guideposts for finding what Aristotle called the "good life"

Martin Seligman

I'm a ruminator, always have been; like a cow chewing her cud I regurgitate scenes and scenarios real or imagined in my mind and masticate them over and over and over again. Montaigne, the great French philosopher, once said, "My life has been full of terrible misfortunes; most of which never happened." That is the story of my life and 'happily ever after' has been the learning to let go of the imagined worries and wars and do the hard work of fighting the real ones.

I know why I worry so much about the people I love. I inherited that from my Mother and from her Mother, Grandma Cole and from Grandma's Mother, Great-Grandma Cole. All of them were strong, hard-working Midwest farmer's daughters. They knew hardship, shortage and loneliness out there on the expansive plains.

Great-Grandma came to Nebraska in a covered wagon when she was five years old and I remember when she was ninety-five she picked up a five foot long Bull snake on her front porch and killed it by swinging it like a lasso over her head and rapping it against the side of the house.

She lived all her life in a little wooden farmhouse seven miles from the nearest paved road.

The farmstead still had a 'working' outhouse around back. A little wooden box with a hole in the ground that contained a hundred years of shit—a hundred years of intimate moments squatting over a hole dug in the frozen ground.

'Shit', that was great-grandma's own word—the one mild profanity allowed by a strongly Christian family. They lived a life surrounded by the substance because cattle create it in enormous quantities and the farmers had to shovel it and trudge through it and wash it off their boots at the end of the day.

I learned from those great matriarchs that love is fretting and worrying incessantly about those you care about and also that the word 'shit' is okay to use in daily conversation.

Self-absorption often equals pain, emotional or physical. Anxiety is self-absorbedly worrying about the future. Defining yourself as an individual is also relevant to the question of mental health versus mental illness. Often, we forget who we are. Have you ever gone against your personal values and done something that you regretted later? Maybe you stole that Percocet from that dying cancer patient and felt guilty later? You forgot who you were. This ambiguity about our self-identity can lead to pain. We see this exaggerated in psychiatric facilities everywhere. Boundaries get blurred. That line-in-the-sand between who I am versus who you are gets confused. The technical term for this is transference and countertransference.

I remember a skin-head, neo-nazi medical doctor (for real, he wasn't delusional) who hated all gays, black people, Jews,... pretty much everyone who wasn't white and bald like he was. He did not have any idea who he was. This is one of the reasons that he hated everyone else, he did not know where he ended and some other person began. While in a locked mental facility, he became friends with a homosexual man. They got to know each other very well, and soon his boundaries between 'SELF' and 'OTHER' were blurred. He decided that he was gay. He realized that everything that he hated before was because he was unsure of his own sexuality. The two men's individual identities became enmeshed. Skin-headed man did not know where he began and his friend ended.

After his friend was discharged, he started to unblur. He became un-enmeshed. He realized what had happened (because this had happened before) and he became violently angry. He even called the other man at home and threatened to kill him. Not really threatened, actually, he made concrete plans on how and when he would murder his ex-friend. Boundary blurring leads to pain. As nurses, sometimes we blur the professional boundaries between us and them. We feel sorry for our patients. We grieve with them. Most of this is normal. Sometimes we agonize over a pediatric death and drink ourselves to sleep as a salve for the pain. Our boundaries have blurred between calloused professional to sensitive friend.

Mindfulness is a simple technique that can help us in TWO ways. We can learn to pay attention to the world around us and not focus internally on our waxing and waning emotional condition. We become less self-absorbed, less contracted inward. We can learn to be mindful of our internal state and realize when our emotions are working against us and self-identity becomes blurred.

A simple way to practice mindfulness is to take a step back from ourselves mentally and watch our thoughts roll by like cars on a freeway. Mindfulness allows us to watch without criticizing or making judgments. Oops, a disgustingly perverted thought slipped by? Oh well, never mind. Just watch it like it's captured in a helium balloon as it floats out of your head and into the sky. If the thought really disturbs you, take a mental pin out and *pop* it as it floats away, and then move on to the next thought.

You can learn to change the way you think about things, and you can also change you basic values and beliefs. And when you do you will often experience profound and lasting changes in your mood, outlook, and productivity.

David Burns.

Truly, thoughts are things—and powerful things when they are mixed with definiteness of purpose, persistence, and a burning desire for their translation into riches or other material objects

Napolean Hill

Though he should conquer a thousand men in the battlefield a thousand times, yet he, indeed, who would conquer himself is the noblest victor.

DHAMMAPADA 103—The Teachings of Buddha.

Another simple practice of mindfulness is to simply pause, take 3 slow breaths, and DON'T THINK any thoughts at all. Focus on the breath. In and out... In this way, we learn to modulate our sympathetic nervous system and our fight-or-flight response. Fight or flight often makes us feel miserable. Our brains CAN NOT tell the difference between real or imagined threats.

This is important, so I will repeat it: physiologically, our body and brain cannot tell the difference between REAL or IMAGINED threats.

We have all seen people (or done it ourselves) react emotionally to a perceived insult with extreme rage. For example, we may have been cut off in traffic by a lousy driver. This causes our limbic system to tell our sympathetic nervous system that we are in a life or death situation. We yell, cuss, and stick out our middle finger. We rant and rave. Our body is reacting to a life-or-death situation when all that really happened is some man could not multi-task his texting and driving.

No one died and no one was even close to death, but our brain and body could not realize that it was perfectly safe.

Have you ever had a terrible nightmare and woke up drenched in sweat with a pounding heart rate? I often have this terrifying nightmare in which an oversized rubber duck is slowly bouncing it's way toward me. One slow water-sodden earth-shaking bounce after another. The dream wouldn't be so scary if I wasn't mired in a foot of freshly poured melted marshmallows and the duck is holding (in it's nonexistent hands), a S'mores skewer with which to skewer me.

Ever had that dream where you are eating a giant marshmallow, only to wake up and your pillow is gone?

Anyway, after a nightmare, our sympathetic nervous is all revved up, ready to respond to a life-or-death situation. Nothing real ever happened. Symptoms of a sympathetic nervous response included all of the things we see in a real life-threatening situation: increased heart rate, increased blood pressure, diaphoresis, rapid breathing, and dilated pupils.

.

"A beggar had been sitting by the side of a road for over thirty years. One day a stranger walked by. "Spare some change?" mumbled the beggar, mechanically holding out his old baseball cap. "I have nothing to give you" said the stranger. Then he asked: "What's that you are sitting on?" "Nothing" replied the beggar "Just an old box. I have been sitting on it for as long as I can remember." "Ever looked inside?" asked the stranger. "No" said the beggar. "What's the point? There's nothing in there." "Have a look inside" insisted the stranger. The beggar managed to pry open the lid. With astonishment, disbelief, and elation, he saw that the box was filled with gold....

Eckart Tolle

Why ER?

I worked in the ICU right after I graduated from RN school and I was just terrified of floating to the ER. I remember dreading the nights when I had to carry the ER trauma pager. The ICU always supplied a nurse to respond to full trauma codes and I never really wanted to be the guy to go, I was just too scared. I had been a nurse maybe six months and I simply was not confident in my skills yet.

One night, my buddy Justin was carrying the trauma pager and the call came that there was a gunshot wound coming in. Apparently a guy had gotten shot in front of a local bar. The patient took multiple .45 slugs and was coding: which means he was dead and CPR was in progress. *Whew*—glad it was Justin and not me responding.

Justin raced down to the ER all excited to help because he was already a trauma-junkie and I felt an immense sense of relief that it wasn't me heading down. I tended to my stable vent patients; you know, doing the ICU stuff like repositioning and fluffing pillows. Then a second call came that there was another gunshot victim coming in and they needed another ICU nurse.

Crap!

No, literally, I felt like crapping my Joe Boxer's. As I started down the back hallways to the ER, I wondered what would happen to me if I really dropped a load in my shorts. How could I cover that up? Especially when I was expected down in the ER stat?

You know, that happened to me once at an inner-city daycare that my parents had sent me to shortly after we moved from redneck Norman, Oklahoma to the metropolis of Los Angeles, California. I was like 7, and I was too scared to go ask the really, really big and fat, impossibly loud, sarcastically-bullheaded recess lady if I could go inside to go potty. As I remember it (and I admit, my perceptions of the past are sometimes flawed) she was like a giant Buddha—but not like the smiley fat guy you see in statue form at Chinese restaurants. She was like a cross between Satan and Buddha: she could burn you in the fires of hell, but she didn't have horns, just fat rolls.

Anyway, I really had to shit and I was gophering it and then *whoops* I mildly crapped my pants and all the other kids laughed at me and my big sister stuck up for me saying, "he just slipped in some mud" and those mean kids demanded that we show them exactly where the mud was that I slipped in. "Sure is some stinky mud" they said and laughed and I felt really yucky inside. I felt like I wanted to crawl in a hole somewhere and never come out. Well, maybe come out for like a peanut butter and jelly sandwich or some orange pop or something, but only occasionally just so I wouldn't starve to death. Then I would go back in my hole and tell the world to fuck themselves.

Geez, that was off-subject tangent, but you can see how this added threat of crapping my pants rocketed my anxiety higher than if the gunman gangster walked into the ER to finish the job.

Anyway, that's how I felt that day as I imagined this big scenario in which I got scared of a trauma patient and I worried that I might accidentally kill the patient, or inadvertently stick a needle in a coworker and infect them with HIV, or forget a drug dose, or crap my pants and the other nurses would all laugh at me! I'm joking a bit for the sake of this story, but you get the idea: I was terrified.

I was able to hold it in and respond to the trauma code and everything went fine and the first thing they asked me to do was take over chest compressions. I remember pushing on his chest, my hands positioned on his sternum between two gaping bullet holes. Every time I pushed, blood oozed out of the holes and trickled toward my gloved hands. I looked at his face while continuing compressions and noticed that he was a young, handsome Hispanic male with his jaw flayed open from the path of a bullet. I had never seen a jaw bone before.

I did adequate chest compressions and one of the other male nurses forgot to connect the chest tube to the tubing and 1500 cc's of blood spilled all over the floor and blood poured all over Dr. Eureka's brand new tennis shoes and he laughed and showed me his shoes later and said, "I guess that's the last time I wear brand new shoes to the ER, I must have jinxed my night!" I was surprised at how friendly and funny and matter-of-fact everyone working in the ER was. One of my nursing school classmates was the paramedic who brought the patient in so I already knew some of the friendly people and the patient died and 50 angry gangster wannabe juveniles mobbed the helicopter pad and the entire night-shift police force had to provide crowd control just for the Life-flight to land and I realized that everything was exponentially more super-cool in the ER.

That's when I realized the next big challenge in my life was to become a trauma junkie.

But it will make solid sense to you, I think. It will slice through a lot of the excuses you've been feeding yourself (and others have been feeding you) for failing to live up to your potential ..... The most accurate thing that can be said about me is that I am "changeable". So are you. ..... from situation to situation, from minute to minute, and even within the same time period and place we have multiple identities that shift and flow in a complex dance of variation. ...... We cannot help but change. This book is about choosing the direction of your changingness and acting upon your choice.

David Reynolds

Why Psychology?

In the year 2000, the American Psychologist devoted its entire millennial issue to the new field of positive psychology. Positive psychology is the study of positive emotion, positive character, and positive institutions (Seligman & Csikszentmihalyi, 2000). The research from positive psychology is hoping to supplement what is already known about the dark side of human suffering (Seligman, Steen, Park, & Peterson, 2005). It makes intuitive sense to me that psychology should have a more balanced approach to understanding human experience: that is, understanding both suffering and happiness.

I was first exposed to psychology through Csikszentmihalyi when I found his book Flow on my father's bookshelf over 15 years ago. The ideas that I read in Flow fascinated me and I still have that old dog-eared paperback on my own shelf. Flow is the mental state of being fully focused and energetically immersed in an activity—it is the psychology of optimal experience (Csikszentmihalyi, 1991). The idea that an individual could proactively pursue optimal experience and a state of flow was revolutionary to me and I still think about it often.

I moved on to read Seligman's Learned Optimism when a friend of mine, a police officer, recommended it. This police officer told me that the book had saved his life at a point when he had been severely depressed. He had found himself sitting on the railroad tracks in his police cruiser with a coal train racing toward his car. He could not find the will or reason to move the car. Learned Optimism taught him cognitive techniques to work on the thinking distortions that had led him to misery and pain (Seligman, 1992). This police officer changed his mind and changed his life, and is now a Christian Licensed Mental Health Professional (LMHP) helping other people do the same.

Seligman's book, Authentic Happiness, introduced me to a bigger world of positive psychology. Because of his writings, I decided to pursue a career in psychology. Psychology became, to me, a scientific path to Aristotle's 'good life'. Seligman teaches simple techniques to increase a person's happiness-quotient such as keeping a daily gratitude journal (2004). Seligman has proven that it is better to focus on improving strengths rather than trying in vain to fix personal weaknesses (Seligman, 2004). These simple, almost commonsensical interventions have been studied scientifically and statistically validated.

Try this exercise: write down 5 things that you are deeply thankful for first thing in the morning and last thing at night. Do this for fourteen days and your "happiness quotient" will statistically improve. Want to measure your level of happiness and find out what your "signature strengths" are? Go to www.authentichappiness.com and create an account.

Martin Seligman is psychology's most vocal proponent of researching the positive side of life. The opportunity to study Seligman's original research, which was done while he was a Postdoctoral fellow at the University of Pennsylvania, excited me. Learned helplessness is the theory that started the entire positive psychology movement. Seligman states that "this question is of particular interest since learned helplessness has been postulated as underlying human depression" (Hiroto & Seligman, 1975, p.311).

I would like to study learned helplessness in nurses. Positive psychology is, quite possibly, the answer to much of human suffering or, what I like to call 'existential angst' and the answer to some of the healthcare crisis.

So, I left the ICU and asked the ogre of ER to 'bring it on'. Which she did.

At the core of the phenomenon of pessimism is another phenomenon—that of helplessness. Helplessness is the state of affairs in which nothing you choose to do affects what happens to you. ..... there is a vast, unclaimed territory of actions over which we can take control—or cede control to others or to fate

Martin Seligman

Can you imagine the personal and organizational cost of failing to fully engage the passion, talent and intelligence of the workforce?

Stephen Covey

Nursing is a portion of healthcare that is in crisis. Predictions vary, but some experts suggest that the nursing shortage could reach 500,000 to 1 MILLION by 2025. This will be about the time when the "baby boomers" are in full-blown, geriatric incontinence. This will be A LOT of work: I need a nap just thinking about it.

The interesting thing is that

THERE IS NO NURSING SHORTAGE!!!

A full 44% of card-carrying registered nurses are simply not working in nursing.

They wizened up and got out. It seems that real-estate has less back injuries, McDonalds pays about the same and provides free uniforms, and early retirement has less dead people and irate families to deal with.

According to OSHA nursing is the #1 most injured on-the-job of all professions. Typically back injuries, but it is also the #1 MOST ASSAULTED profession (500,000 a year): so a smashed up face is not all that uncommon either.

Another frightening statistic is that about 30% of nurses abuse alcohol or drugs.

Six to eight percent are impaired ON-THE-JOB!

This means that out of one-million nurses 940,000 nurses are NOT drunk at work!

Nursing is a segment of the healthcare industry that is in crisis!

The American Association of Colleges of Nursing (Rosseter, 2008), are the ones who say the shortage of registered nurses (RNs) in the U.S. could reach between 500,000 and one million by 2025. A large part of the problem is the job stress and dissatisfaction that RNs experience. Rosseter (2008) reports that 37% of RNs that are currently employed are unhappy enough that they are considering changing jobs. Another problem is the high rate of chemical abuse and misconduct by nurses. There is no research available to tell if this is a symptom or a cause of nursing burnout.

The number and percentage of nurses with sanctions imposed by state boards of nursing have risen in the last decade, with male nurses being disciplined at a rate twice that of female nurses (Zhong, 2009). In 2005, the American Journal of Nursing suggested that it may simply be Post-traumatic stress disorder (PTSD) from on-the-job trauma that is driving nurses from the profession (Schwarz, 2005).

Personality characteristics like high versus low emotional reactivity possibly play a role in nursing burnout and job stress. Attributional style or explanatory style indicates how people explain to themselves why they experience a particular event as either positive or negative (Seligman, 1992). Learned helplessness is the psychological state in which an individual comes to believe that they are powerless to influence the outcome of a situation.

Hiroto and Seligman (1975) published seminal research demonstrating that learned helplessness can be produced in humans. The authors define learned helplessness as "inescapable aversive events presented to animals or to men" (Hiroto & Seligman, 1975, p.311) that results in profound interference with later learning. Learned helplessness looks like *depression*.

Learned helplessness is the concept describing how individuals come to believe that their actions or behaviors are unable to impact the outcome or result of life strivings and goals (Schepman & Richmond, 2003). It is difficult to conceptualize this problem because of a lack of research regarding the role of individual personality characteristics in learned helplessness. Research shows that repeated feedback with no predictable relationship to the actions of the individual tends to cause learned helplessness (Seligman, 1992). Why do some individuals become helpless in situations that do not impact other individuals negatively?

According to Hiroto and Seligman (1975) a recognizable debilitation is produced by uncontrollable events and can be generalized across different motivations and tasks performed. This means that not only does work suffer; but nurses often take shit home with them!!! (this is called secondary trauma or vicarious trauma).

Empowerment, the power to influence outcomes, is the opposite of helplessness.

An empowered worker believes in his or her own personal efficacy.

Efficacy means we have the capacity; the ability, to change our environment (our reality)!

Immanuel Kant (1724-1804) goes so far as to suggest that we have personally created the reality in which we now live! There is always something that we can change or influence in our life. Relationships, jobs, living situations can all be changed. More importantly our *response* to situations can be changed.

Brandstatter (2001), a psychologist from the University of Linz in Austria has done extensive research comparing personality type and emotional responses of people to everyday life situations. He is interested in finding out how different people react differently to various situations. An important part of different reactions are individual's personality types, particularly their temperament. He states that, "temperament is understood as a person's mostly inherited general "reactivity" (sensitivity) of the nervous and endocrine system" (Brandstatter, 2001).

Learned helplessness leads to nursing burnout and might be due to individual personality and temperament. My idea is that high-emotionally reactive individuals react poorly to emotional invalidation (invalidation is nursing in a nutshell!) These emotionally invalidating environments create internal pain that is magnified in intensity by learned helplessness.

Larsen and Buss (2008) discuss how people are wired differently in terms of emotional reactivity. They did brain scans of people with many different personality types. This new research using functional magnetic resonance imaging (fMRI) has shown a difference between how negative or positive visual images were processed differently by people. In particular, individuals with the personality characteristic of high levels of neuroticism or high levels of extraversion.

Canli, et al. (2001), showed neuroticism correlated with increased frontal brain activation to negative-emotion producing images whereas extraversion correlated with increased frontal brain activation to positive-emotion producing images. Simply put, some people have stronger emotional responses to negative events. Other people have stronger emotional reactions to positive events. 'Glass-half-empty' pessimism seems to be somewhat of an inherited, brain-based personality trait.

For this next exercise, I'm going to share some ER stories. Try and think about these experiences, which might be very similar to ones that you have had in your work; think about the stories from the paradigm of high reactivity and learned helplessness. How would people who are genetically wired differently react in different ways? What are some aspects of this reality that could be influenced or changed by the nurse? Learned helplessness comes from the inability to change the outcome of a situation. Can nurses change the outcome? If not changing the situation, can nurses change their own personal response to the situation?

7   Some ER Stories

"You ask, "Why do many adversities come to good men?" No evil can befall a good man; opposites do not mingle. Just as the countless rivers, the vast fall of rain from the sky, and the huge volume of mineral springs do not change the taste of the sea, do not even modify it, so the assaults of adversity do not weaken the spirit of a brave man"

Seneca

Something I've learned about emergency room nursing is that just when you think things could not possibly get any worse, they do. Just when you think you have seen it all, something unimaginably bizarre walks through the door. "Walk", is a word that I used just now as a literary device because it sounded nice. These people usually don't walk. They are often carried through the door by loved ones, gangsters, wheel-chairs, wheel-barrows, and gurneys.

Here's a snapshot of a couple of patients amongst hundreds, on one day amidst dozens, in one month (November) amongst a decade or so. I guess there's nothing dramatic here, they just happened early in my ER career, so they stayed with me. These stories are routine, non-dramatic experiences and hopefully set the tone for the rest of the book. I guess it's my book; I can digress and share what I feel like when I feel like digressing and sharing. I don't consider these bad scenarios, they were actually quite fun: the kind of patients ER nurses love. Later, I'll talk about some of the patients that we hate.

==

It was a cold, rainy Saturday night in November and a quiet one for the Emergency Room. Not many of the normal guests-with their psychological problems manifested as various psychosomatic body aches-were interested in braving the rain that fell in sheets outside. The modern ER trades in a currency of Morphine, pregnancy tests, and work notes, but these economic incentives were not enough to woo the weekend clientele who would have to wade through the ice-cold curbside puddles to the brightly lit emergency room entrance.

The quiet inside the emergency room was broken suddenly by the familiar, monotone female voice of the police dispatcher on the scanner. The cheap Radio Shack scanner is the valued ear into the city that lies beyond the walls of the hospital. Nurses often plan and triage their care by knowing how many ambulances are out on calls and reserving a bed or two if critically ill patients are on their way in.

This time the dispatcher announced that "RP (reporting party) reports over a dozen gunshots fired in the vicinity of Lyon's Club Park" and then, "10-8 (for your information), three more calls from other houses are calling with similar reports."

The new guy's voice, possibly a little too measured and calm—an over-compensation for this sudden mid-donut adrenaline rush—came over the scanner politely requesting all available units to respond to this potential gun battle. Lyon's Club Park is traditionally where the young gangs of this small Midwestern city like to settle their disputes.

I was new to the ER at this time and I was following Chad, an experienced E.R. nurse to learn the ropes of working in the Emergency Room. Chad listened to the scanner and walked slowly to the trauma room and was already putting his full trauma gear on. To protect against any potential blood baths the nurses like to wear waterproof shoe-covers, gowns, gloves, masks and goggles.

Chad said to me, "You'd better get your stuff on; these usually come quick without a chance to think."

In my own mind I doubted that anything would come of this. Like David Letterman we liked to play the game, "Is This Anything?" and usually we are correct in guessing that no, it isn't anything. I dutifully donned my trauma gear in the off chance that Chad knew something that I didn't.

Seconds later a very excited officer's voice came over the scanner. "I just had a car burn through a stop sign in front of me. They're moving fast down Faidley [avenue]; I think they're heading to the hospital!" He then added, "You'd better let the E.R. nurses know to get ready; I think they're going to have a patient."

I still didn't have my gown quite tied when a handful of dripping wet juveniles came running through the sliding E.R. doors carrying a shot, bleeding friend. Chad pushed a trauma cart out into the waiting room and I helped a shirtless, muscle-bound, brown-skinned, adolescent lay his buddy on the wheeled bed. Before I closed the security doors behind the cart with the dying kid, I could see the lights of a multitude of police cars flashing behind the thick rain through the sliding entrance doors.

==
==

Luke was not sure when the nightmare had started. He wasn't even really sure how old he was. For me, the story all started when the triage nurse came dancing back to the nurse's station, happily waving a chart at me. "Josh! I have a great one for you… 16 years old and totally psychotic" she said, with an F-Y-F smile on her face. 'Fuck-your-friend' is ER-speak for 'I am so glad this is your patient and not mine!'

'Damnit…' I thought to myself as I took the chart. The damnit was the sort of damnit that Jerry Seinfeld might say—just a touch of humorous irony combined with real-life bitterness. The door to the exam area of the ER opened and a handsome young man with dark hair and darker eyes shuffled down the hallway toward me. He was followed by a friendly sheriff from a neighboring small town. The cop had a perplexed and concerned look on his face.

Luke had darting eyes and seemed to be intently conversing with people that I could not see. At the nurse's station (which was halfway down the hall from the exam room where we were going) he turned and stopped to talk in a louder voice. There were doctors and nurses milling around the station, but he did not see them and was not talking to them. Rather, it appeared that he was arguing with someone above their heads and someone who seemed to be some distance away. I said, "Luke, can you come with me?" and he burst into profound sobs, with giant tears running down both cheeks. He slowly turned to follow me with his shuffling gait; all while mumbling and wringing his hands.

Once he was in the exam room, he immediately flopped onto the bed and continued his internal dialogue. I wanted to find out more information and said, "Luke, can I talk to you?" His head was turned away from me toward the wall and he continued mumbling words that I could barely understand. He put his hand up when I spoke as if to say, "excuse me, I am having a conversation here!" pointing toward the blank wall, and then he started sobbing again. It was the kind of sobbing with violent chest heaves, almost like an animal in pain, and the giant tears started again. My first thought as an ER nurse was, 'well, at least he's not too dehydrated'. Tears are a good assessment in that sense—a symptom of adequate fluid intake.

I wasn't sure, but I think I could understand him saying quietly, "but I just don't understand. Why did you take her away from me?" and then... "Where is she?"

"Come back, don't leave, come back..."

The doctor ordered a healthy dose of Haldol, an antipsychotic, while we waited for his lab tests and a urine drug screen to come back. The cop, truly empathetic (he should have been a nurse ;-) said, "Is he on drugs?" I replied, "Maybe, he looks like he could be Robo-tripping, but we need to make sure it is not something worse like a huge cancer pressing on his frontal lobe or something". The cop said, "oh" and then asked, "what's Robo-tripping?"

"It's high amounts of dextromethorphan, the 'DM' in Robitussin DM. They can look like a patient on phencyclidine (PCP) or LSD, with hallucinations and the whole bit".

The officer nodded and with a concerned look, "Is he going to be okay?"

"Well, we need to know his history. Why did you pick him up?"

"Oh, he was in the county jail about 200 miles north for the last two weeks and the sheriff up there asked me to bring him back here because we think this is his hometown".

"When did he start acting this way?"

"The jailer said he has been this way the entire stay. They said that he is just on drugs."

"Oh, so they provide drugs in jail now?"

The nice officer was just doing a favor for the sheriff in another small town and they had worked out a deal to meet halfway so that neither one would have to drive quite so far. I said, "You know, we are going to need to place this kid in emergency protective custody if his lab tests all turn out okay"

"Well, hell, I can't get involved in that! I'll call Bob, the original arresting officer and make him drive down".

The Haldol made Luke pretty sleepy and he passed out for a while. All of his lab tests, head CT, and urine drug screen came back completely normal and eventually the small-town sheriff from 200 miles away in sheep-screwing-somewhere showed up. My initial impression of 'Bob' was a dickhead with small-penis syndrome. "What the hell am I doing here? This asshole does not need to go to a psych-ward, he's on meth! I am taking him back to jail!"

I said, and in hindsight it wasn't the brightest thing to say, "Oh, so you guys provide meth in jail? It's my understanding that he's been locked up for the last two weeks". Red-faced and stammering, "well, what the hell else could it be?"

"He is not on drugs. His drug screen came back completely negative."

By this time Luke was starting to wake up and he was much more lucid. I turned to him, "Luke, how are you doing?"

"I think I feel a lot better!"

"Luke, do you have any medical problems or psychiatric diagnoses?"

"Oh, yeah, I have paranoid schizophrenia… where am I?"

"You are in the emergency department". I pointed to his name band, "Is this you? And did we get your birth date right?"

"Yeah, that's my name… but it says I am 16 years old!" he started crying.

"I can't be 16! Where did all those years go?" big crocodile tears again.

The asshole cop says, "What the fuck is schizophrenia? I have never heard of it. I'm taking him back to jail."

An EPC take hours of paperwork for the officer involved and in Nebraska (which ranks 49th in the nation as far as mental-health resources go) can be nearly impossible.

I said, "have you heard of Alzheimer's disease?"

His face softened, which surprised me, "yes, my Mother has it. She's in a nursing home. She has lost all memory and doesn't even recognize me anymore".

I waved toward Luke who was blowing his nose into a tissue that I had offered him. "Luke has a disease that is similar in the sense that he is slowly losing his mind. If we did the right kind of scan on his brain, there is a chance we could even see the part of it that is slowly degrading. Look at him, a good-looking kid with a bright future. This is as good as he is ever going to get and he will probably spend his life in institutions. His brain is not working properly, just like your mom!"

Something dramatic had changed in the officer's demeanor. He stepped outside into the hallway without saying a word to me and dialed a number on his cell-phone. I heard him say, "Yea, I am going to be here the rest of the night. We need to EPC this kid and get him some help."

As it turned out, no facility would take Luke because he was a minor and had no parents that could admit him. The only possible way to get him treated was to make him a ward of the state: something that is impossible at 2am on a Saturday night. Bob spent the next six hours arguing with people from some state department in the capitol and made Luke a ward of the state so that we could send him to an appropriate psychiatric facility. I now consider Bob a small-town hero.

I asked Luke, "When you first came in, what was happening? You were hallucinating, what were you seeing?"

He started to cry a little, "My mom left about six months ago to go back to Texas. She told me that I would be fine here. I thought that I was in a small town in Texas and I would find her and ask her to help me, but then she would disappear again. I almost found her, I almost got to her, but they kept taking her away again."

"What is happening to me? Why can't I find my mother?"

The science of psychology [historically]........ People were assumed to be products of their environment. The prevailing explanation of human action was that people were "pushed" by their internal drives or "pulled" by external events. Though the details of the pushing and pulling depended on the particular theory you happened to hold, in outline all the fashionable theories agreed on this proposition. The Freudians held that unresolved childhood conflicts drove adult behavior. The followers of B.F. Skinner held that behavior was repeated only when reinforced externally.

Martin Seligman

To really succeed in a business or organization, it is sometimes helpful to know what your job is, and whether it involves any duties. Try to find this out in your first couple of weeks by asking around among your coworkers.

Dave Barry

NATURE VERSUS NURTURE:

Marsha Linehan was a brilliant therapist who worked to help highly suicidal clients with serious problems like regular self-mutilation. Most of the clients suffered from Borderline Personality Disorder. Linehan (1993) in her book on the treatment of borderline personality disorder (BPD) suggests that there is interplay between two separate components that cause a problem with personality: high emotional reactivity which is genetically brain-based (i.e. nature) and invalidating circumstances (which are the nurture/environment part of the equation).

Changing the genetics of a nurse's brain is not a feasible workplace intervention. However, changing the environment (i.e. the invalidating circumstance) is a possibility that is within reach of all healthcare organizations. The problem for many nurses, emergency medical personnel, and physicians (if you can call normal personality characteristics a problem) is simply emotional reactivity.

Reactivity has to do with how dramatic and intense emotional reactions are to various situations and stimuli.

I suspect that the idea of emotional reactivity might be eliminating the term "Borderline" in the upcoming DSM-V, which is due to be published (maybe) in 2013. This personality trait is something everyone has: all people are on a continuum from high reactivity to low reactivity. We all fit somewhere on the continuum. Something that I am interested in researching is Emergency room nurse's reactivity and how it correlates with rates of burnout and other forms of dysfuntion. Some people have very profound and intense reactions, others never seem fazed emotionally. It's a bell-curve, so most of the population is somewhere in the middle.

People with high reactivity make great nurses, clergy, teachers—they tend to be empathetic and caring—they want to help people. They also tend to be self-destructive due to the invalidating circumstance of modern healthcare or public service. I have been wondering lately if maybe the genetic, pre-set level of reactivity isn't reset by extremely traumatic events? Our soldiers (heroes) returning from war zones certainly seem to have a revved sympathetic nervous system (can you blame them?)

Marsha Linehan, the founder of Dialectal Behavioral Therapy—which is the only thing that has ever shown any promise in treating personality disorders—argues that any personality disorder (especially borderline) is the function of two things:

1) genetic predisposition to high emotional reactivity and 2) invalidating circumstances.

What is an invalidating circumstance?

Something that makes you feel in a way that you think you are not supposed to feel…

Psychologists once viewed childhood sexual abuse as kind of the gold standard for an invalidating circumstance. Another example would be that "sensitive boy" who cries easily and his gruff, tough father smacks him around and says, "Little boys don't cry, take it like a man!" That little boy is "invalidated" and grows up unable to trust his own internal emotional state. He is a 'high emotional reactive' and apparently he misperceived how he was feeling, and it was violently reinforced for him him that he was not allowed to feel, or trust his emotional state. He can't regulate his emotions because he cannot even accurately tell what those emotions are anymore.

Anyway, there is still a lot about this theory that needs to be studied, refined, changed, or thrown out. However, what I believe is that

The circumstances are not nearly as important as the degree of (genetic) emotional reactivity.

Case in point: I always enjoy talking to ER patients with "post-traumatic stress disorder" about the event that actually caused their PTSD. Sometimes they are fresh from Iraq where they saw buddies blown to bits with roadside bombs. Sometimes, however, the event is something so innocuous as to be quite funny. For example, "I took the driver's license exam and now I have to be heavily medicated for my post-traumatic" or "the SAT test was the cause of my post-traumatic, now I am on disability".

True, these events can be stressful, but most people do just fine getting through them. Actually, the same is true of war and sexual assault and all of the other stressors in life—the high-reactives don't do well and the low-reactives move on apparently unfazed. Extremely high-reactivity can be very frightening. Simple things like reading the newspaper can create profound feelings and reactions to the horrific atrocities committed against humanity every day. Life becomes frightening.

This nightmarish emotional reactivity is modulated by the limbic system in the brain, and the sympathetic nervous system (SNS) that has effects throughout the entire body. I think it is important for nurses to learn how to assess SNS symptoms (like dilated pupils, increased HR, sweating, etc. etc.) and find out more about modulating its sinister effects.

The limbic system in the brain cannot tell the difference between a real or imagined event. The event could be disastrous, such as a massive earthquake, or the event could be the terror of watching the Smurf's movie trailer. The body reacts the same way.

For the next exercise, I will tell you another critical-care story from last summer. Try to imagine that you react emotionally to some of these events. Are they invalidating? Do you find yourself feeling a certain way that "professionals" are not supposed to feel? Or are nurses perfectly sane and working in insane situations…

I suggest that circumstances are only part of the problem. In fact, I suspect these situations really have nothing to do with the woes that plague the nursing profession, it is a genetic predisposition to empathetic sensitivity that causes many nursing (hero) problems! It's the old nature versus nurture debate (genetics vs environment/reactivity vs invalidation) it's all the same question: what came first the chicken of emotional reactivity or the egg of invalidating circumstances?

Genetics is the cause of the high emotional reactivity that allows nurses to empathetically care about people. But, with some invalidating circumstances thrown in for good measure; genetics can also lead to a nurse's self-destruction. For nurses a symptom of the problem is the 30% (or higher) abusing drugs and alcohol! Almost 10% of nurses are working impaired!

"Most men lead lives of quiet desperation and go to the grave with their song still in them"

Henry David Thoreau.

"The power to live with joy and victory, is available to you and me. This power can lead you to a solution to your problems, help you to meet your difficulties successfully and fill your heart with peace and contentment."

Norman Vincent Peale

"You are fooled by your mind into believing there is tomorrow, so you may waste today." Ishin Yoshimoto

Here's a story that happened to me the other day in ICU. It impacted me because, for some reason, it bothers me a bit when one of my patient's dies. Even though all people tend to do that at some point in their life, and I work in critical-care, the part of the hospital where people tend to die—so, I don't know why it occasionally bothers me...

Ultra Concentrated Joy

I am doing the dishes by hand again. Our dishwasher died a while ago, and we have not quite gotten around to buying a new one to replace it. It's okay though, we don't tend to generate a completely unmanageable volume of dirty dishes per day, and sometimes it's relaxing to do them by hand. I often think about my last day at work as I use the raspy side of the sponge to get the little bits of hardened food off the plates under the steaming hot water in the dishpan.

Yesterday, I worked in the Intensive Care Unit at a hospital an hour away, down highway 30. It's a progressive setting that deals with a widely diversified patient population: anything from open-heart surgery patients, to neurologic catastrophes. Traumatic 'train-wreck' patients or patients literally hit by a train. The busiest stretch of industrial railway in the world runs around this town. The hospital has a flight crew that brings the sickest patients from a two-hundred mile radius to receive the best medical care that the world has to offer.

My hands are starting to get white and wrinkly from the hot dishwater and the pile on the dirty side is slowly getting smaller as I pile them ever higher in the clean dish drainer. The cheap Radio Shack under-the-counter radio is playing the newest song by Coldplay. Chris Martin's baritone crooning an emotional sentiment in his easily recognizable falsetto:

"For some reason I cannot explain I know Saint Peter will call my name"

Ronald was one of my patients yesterday. He had end-stage everything. I think the list included chronic obstructive pulmonary disease, cardiomyopathy, sleep apnea, asthma, emphysema, diabetes, coronary artery disease... You get the idea. Essentially, you name it; he had it.

I spent the day talking to the family about the dreary outlook of his prognosis. One of the doctor's had told me point blank: "He'll be dead today, absolutely no doubt." The cardiologist, an elderly, soft-spoken Indian man was a little more lenient: "I'll give him a 50/50 chance of dying today; he might just make it until tomorrow." The oldest daughter; a happy, zaftig professional in her later forties was the spokesperson for the family and she said, "You know, after his second child—his youngest son—committed suicide at the age of 21, he died."

"He hasn't been alive inside for at least five years..." I was poignantly aware of how she emphasized the word 'inside'. She nodded at the patient's wife who was now crying softly to herself in a corner of the room. "Neither one of them has really been alive since then; none of us," She gestured toward me, "can really imagine what it is like to lose a child, let alone two."

66

I continue to scrub the skillet that had been used this morning to cook scrambled eggs for breakfast. Coldplay is still singing: "For some reason I cannot explain..." The bottle of dishwashing soap mocks me: 'Ultra-Concentrated Joy', it says on the fucking label over the fluorescent pink, fragrant liquid.

"How the hell do we get some of that?" I ask aloud to the silence of the whitewashed kitchen cabinetry.

Ronald had actually awoken after we withdrew all the invasive treatments that had been ventilating his lungs and keeping him alive. He hugged each of his children and told them he loved them. There was a lot of crying and talking and hugging, so I kept my distance for a while. Later, I tried to talk to him but he seemed a little confused and I said as much to his wife. She said, "Oh don't mind that. He's been that way for at least the last two years. In fact, he has been hallucinating and talking to dead people for a long time. Sometimes just one or two, but sometimes it seems like the whole living room is full of dead people and he just carries on talking to all of them as if they are really there. He gets upset with me because I can't see or hear them at all."

"Really?" I had asked, "Are these dead people... uh... people that he knows?"

"Oh, yeah, his Dad and Mom and his two kids that have passed. You know, his dad lived with us until he was ninety-four. His dad had planted a big garden in our backyard and was out working in the garden when he died. It was actually a kind of beautiful death: he was reaching into his pocket for a tool and just fell over dead, his hand still in his pocket."

She paused to wipe a tear from her cheek with a weathered hand. "He died doing what he loved to do and he didn't suffer at all. Unlike this..." she waved her hand toward her husband who was now lying in bed breathing fast and hard. His body shook with each breath, like a coal train accelerating past the town and into the next state to deliver its load.

"My husband still took it hard though, I think because we didn't find the body for a couple of days since we were out of town."

I'm emptying the dishpan of dirty, soapy water down the sink drain and preparing to fill it again with new, hot water. Reaching for the Ultra-Concentrated Joy, I again wipe my hot cheeks on my shirtsleeve to eliminate the moisture that is accumulating there. There are only the cups left to wash and the radio has moved on from Coldplay. I wish that I had a few more dishes left to do, because I just don't quite have shit figured out just yet.

Ronald's family had stepped out for a quick bite to eat and he sat up in bed and asked me to help him slide up in bed a little farther. The bend in the bed was folding him just a little wrong and had created a backache for him. Before I could get to the side of the bed, he had pulled his legs up and tried to push himself up, using his arms against the side rails for extra assistance. This energy expenditure was entirely more than his failing heart and lungs could handle and his face turned a bright purple; the cyanotic color of a dead man. "Ronald, stop struggling," I said, putting my hand on his shoulder, but, it was too late; Ronald did not look up at me, or even take another breath. Ronald was dead.

10   Invalidating Circumstances

"As human beings we have the potential to disentangle ourselves from old habits, and the potential to love and care about each other. We have the capacity to wake up and live consciously, but, you may have noticed, we also have a strong inclination to stay asleep. It's as if we are always at a crossroad, continuously choosing which way to go. Moment by moment we can choose to go toward further clarity and happiness or toward confusion and pain."

Pema Chodron

Invalidating Circumstances

When was the last time you did CPR? How about CPR on a 4 year old, or a two-year old, or the last time you crawled over a mangled (obviously dead) 19 year-old girl in an upside-down car in a ditch so that you could open the airway of her partially alive friend? Most paramedics have, that's why we call them heroes.

That's also why fire-fighters, like nurses, have such a high rate of divorce, dysfunction, and self-destruction.

By the way, no one really ever survives CPR and as a nurse, you know how it feels to occasionally (my wife—an ICU nurse—says "always") be blamed for the poor outcomes. Apparently, science has not yet perfected the art of bringing people back from the dead.

I remember a local firefighter who came in dead. Pretty hard when every-time you do a chest compression blood oozes from every orifice due to DIC (disseminated intravascular coagulation). The mattress had to be thrown away like a bloody sponge. Harder still were his young children screaming and crying at the bedside saying "please save my daddy" over and over again. Surprisingly, the nurses and doctors could not save daddy, and they felt invalidated—feeling emotions that a "professional" is not supposed to feel.

Invalidating circumstances? I think so. Does the emotional reactivity lead to an irrational fear of abandonment? Hmm... probably. These are all symptoms of 'borderline personality disorder' which thankfully won't exist in a couple of years, they are also symptoms of "secondary trauma" the idea that emergency workers take shit home with them.

Genetically predispositioned high reactives take more shit home than low reactives. High reactives write songs and novels to try and put words to their catastrophic emotions (the corpus-callosum forges new synapses when a person writes—effectively connecting the right brain and left brain—similar to EMDR, a treatment for PTSD).

Where did this emotional reactivity come from? It's pretty hard to do a genogram of mental-health disorders (though a fascinating read is Kay Redfield Jamison's "Touched with Fire" which analyzes the link between manic-depression (in my opinion, old terminology for extreme reactivity) and many great artists & poets—complete with genograms (i.e. family tree) of mental disorders. It's hard to analyze family history because we (society in general) don't talk about mental-health.

Most likely, high reactivity (which is not a problem, just life) is inherited. High reactives are notorious for closing down, not talking, sulking. It's simply the deep pit of emotional despair (modulated by the limbic system) makes talking very painful and difficult.

The body is telling the brain that they are in a life-or-death crisis.

Dale Carnegie talked about this in his book "How to Win Friends and Influence People". According to Dale, when we are criticized it feels like a life-or-death situation. Interestingly, Buddhism has discussed and dealt with this problem for thousands of years. When we are contracted (closed down) emotionally we tend to lash out as a protective mechanism—borderlines even get a little paranoid, constantly worrying about perceived emotional threats: a lover cheating or leaving, friends laughing at them behind their back, or co-workers doubting their ability as nurses.

Mindfulness meditation is one of the few things other than drugs and alcohol that helps high reactives feel better (medications esp. Lamictal or lithium or Abilify help too...)

You combine the genetics for emotional dysregulation (from high reactivity) with the tendency toward Anxiety disorder or Bipolar disorder (both DSM diagnoses) and the impulsivity and grandiosity that are symptoms of bipolar, and you've got a mental-health emergency in the making.

You also have the potential for some brilliant creativity and powerfully, impactful lives.

Important: all this stuff is not a big deal! Just life. Life needs dealt with occasionally and we call this an "existential crisis".

For some time, I have been collecting song lyrics that reference a high emotional reactivity. In my conferences we play, "name that mental illness" and watch You Tube videos of these songs. I love "hard rock" because rather than being tough guys as they try to appear, heavy-metal bands are often extreme emotional reactives. For a great discussion of Bipolar Disorder listen to anything by "Blue October". For Borderline, anything by the pop-star "Pink". Check out "Tears Don't Fall..." by Bullet for My Valentine:

"There's always something different going wrong

The path I walk's in the wrong direction

There's always someone fucking hanging on

Can anybody help me make things better?

Your tears don't fall

They crash around me"

"Tears that crash..." implies that the singer feels things a little more intensely than most people. Most tears don't actually crash... you get the point...

11   We Have a Choice

Nevertheless, we have a responsibility to think bigger than that these days. If spiritual practice is relaxing, if it gives us some peace of mind, that's great—but is this personal satisfaction helping us to address what's happening in the world? The main question is, are we living in a way that adds further aggression and self-centeredness to the mix, or are we adding some much-needed sanity?

Pema Chodron

Choice

I am not certain from a theological perspective if free will exists but, like William James

I have decided that my first act of free will is to believe in free will.

What a sense of empowerment to know that we have the power to choose and the power to create the reality in which we live.

Immanuel Kant (1724-1804) first suggested that we live in a reality that we have personally created, primarily through our own perceptions.

Stephen Covey, an organizational behaviorist wrote in his famous book, "The Seven Habits of Highly Effective People":

"between stimulus and response is our greatest power—the freedom to choose" (1989, p.70).

Nurses, paramedics, physicians, can choose their response to any situation or stimuli…

How can we show people that they have a choice and the tools needed to maximize their effectiveness?

Cognitive-behavioral therapy (CBT) is a clinical psychology tool that can be translated into the workplace, where it is called cognitive-behavioral modification (CBM) (Boan, 2006).

Aaron T. Beck in his book Cognitive Therapy and the Emotional Disorders (1976) explains that it is at the point between stimulus and response where we choose our reaction. This point is also where most of the stress-induced psychological injuries occur.

In cognitive-behavior modification (CBM), "change starts with observing behavior through heightened awareness and deliberate attention" (Boan, 2006, p.53). CBM uses the idea that individuals live in a reality that they have personally created and this reality may or may not be an accurate representation of the external environment (Boan, 2006).

The narrative metaphors (which are the words and stories people use to describe their reality) gives clues to possible inaccurate interpretations of reality. We can use the words, images, emotions, and stories as a starting point to change. Imagine you are an emotionally high reactive nurse. Many nurses are, which is what makes them caring and empathetic for people going through hard times. How will you deal with invalidating circumstances?

12   Some Sad Stories. i.e. What Nurses Do For A Living

This story happened about the same time as the story about that guy in the ICU who died. I've told this second story once in front of an audience (kind of a keynote address) at a large University for brand-new nursing graduates and I must have emphasized the right places at the right times because everyone was crying and we had to break early for lunch to try and find more Kleenex boxes. I felt like I had unfairly manipulated their emotions. I'll try not to do that again...

The fact of that matter is that this story is fairly representative of what ER nurses do for a living. It's one of the crappy patients that no one likes to get, but we do our job, and the sun still rises the next morning for the ER staff (though, I suspect, not necessarily for the family involved). The infant mortality rate in the United States is still fairly high, which means that this happens with some frequency in ERs throughout the country and the world (CDC, 2007). I recently had a nursing student of mine experience almost an identical scenario and as she wrote about it in her journal, the stories were strikingly similar.

Last summer keynote address at the University

I'd like to tell you about my last summer in the ER, because it was a supremely shitty one!

I went to my first Critical Incident Stress Debriefing (CISD). A CISD is a gathering of everyone involved in a shitty, tragic, life-altering incident. A licensed mental health professional oversees the meeting and we all go around the table and talk about what we experienced.

At first, I refused to go, I've never been to one and I've seen a lot of shit, why should I start going now?

And fuck if I'm going to talk!

I'm an ER nurse, goddammit, and we are made out of... well, the metaphor escapes me, but we're made out of something really tough and strong. Like the space shuttle, you know, something able to withstand a lot of shit...

Well, two of my friends, Brenda and Louise pimped me just a little and said, "Josh, it's not for you! It's for all those other people who need your brilliant insight!"

Oh, since they stroked my ego in that way I'm more than willing to go. I'm always willing to save the world and try to fix people. Little did I know I'd be the one bawling like a baby… (Just kidding, I never cry until later at home alone when the vodka has kicked in…) but I guess the CISD did help, maybe a bit, who knows: whatever, nevermind….

The reason that we had a CISD was that a 4-year old was shot and killed. Dr. Flobot and Dr. Gunner did an open thoracotomy to try and release the pressure on the little dude's heart.

That's means that they cut his chest open right in ER trauma 2. He had cardiac tamponade, which is a rapid accumulation of blood in the pericardial sac. The sac is non-distensible which means that the blood starts to compress the heart and prevent it from beating. He was dead when he arrived to our ER and had probably been dead for 45 minutes or so.

The little guy got shot playing outside with his big brother. They had found a high powered BB gun in the neighbor's dumpster and he was able to walk to the house and go inside and tell the babysitter "hey, you know… I got shot right here (rubbing his chest)and it kinda hurts…"

and then he died.

We may have been able to make a difference with the open thoracotomy except for the fact that he had been down for so long.

I tried to take a clinical interest in the pathology of the subject matter (anything to try and cope).

Thinking that it would help, I leaned over like a fucking scientist with the four Doctors and looked at his heart, "uh-huh, here's where the bullet went in. Looks like the left atria… yep…"

Just like dissecting a fucking cow's heart in Anatomy and Physiology class at the university.

Gary Reinkely, a buddy of mine who went to RN school with me (while simultaneously completing his paramedic degree) noticed my beet-red face and said, "Josh, you okay". I nodded him off, like a guy is supposed to do and grunted, "yeah…" But I'm on the phone home to check on my own little blond haired, blue eyed 4-year old. I'm thinking how his devilishly, mischievous smile is all I really want to see right now.

Amy is used to those calls:

"Hi, is Nate okay?"

"yeah…"

"what's he doing?"

"Just watching Sesame Street. Is something wrong?"

"No… Bye…"

And I hang up and start finishing my charting, printing off the last 6 seconds of asystole and taping the rhythm strips to a piece of paper, careful to avoid eye contact with anyone and it's okay because they are all carefully avoiding eye contact with me.

The only problem was that the image of that little dude's heart haunted me in my sleep for a very long time. I remember sobbing in bed as Amy looked concerned (she always knows enough to not say anything) and I kept saying: "why did I have to look at his heart? If only I hadn't seen his dead, baby heart…"

So the next day, I show up for this CISD and I walk right into another tragic event: an 18 month old strangled to death on some mini-blind cords…

I'm like, "why is that baby in exam room-one so quiet and bluish… shit, it looks like she's dead!" and Kelly, one of the hot blond nurses that I've been accused of sleeping with, though I never did because things just never worked out that way, said, "Yeah, she's dead…it's been a shitty day."

84

And I said, "oh".

God, I felt so bad for Gary Reinkely, he was with me yesterday as a nurse and today he was the paramedic who ran from the house carrying this other blue baby, while the 10 year old baby-sitter screamed uncontrollably. Laura and Kelly did CPR on that little body for an hour after she came in and now they've got two CISDs to talk about.

About a week later, we were working and it was a very rare moment in the ER where we had absolutely zero patients. Someone, I think Christy Marocick, had made yummy brownies and all of us nurses were in the breakroom chowing down!

Because that's what nurses do... we fucking chow down chocolate and anything else that we can get our hands on. You see, when we're working, diets don't fucking matter. Bring on the Krispy Kremes and Starbucks double-fudge chocolate-latte' lard-ass ice-cream—we're working damnit!

We'd be drinking if wasn't for the damn, random urine drug screens (though I always suspected that Christy Marocick had her thermos full of her own special little motivational juice ;-)

Anyway, it was in mid-brownie-bite that I heard the front-desk registrar scream. It was a real scream too, not a 'maybe I should talk louder' scream. But a blood-curdling scream.

"I need a nurse out here!"

Fuck.

We ran out into the hallway and there was this ashen-faced Dad, holding his white baby. Dad had a question mark face. Babies aren't supposed to be white. They're supposed to be all fat and pink and bubbly with innocent joy. The dad had the kind of perplexed face that was terrified because it already knew the truth, and I scooped up the baby: not because I wanted to, damn it, there was just nobody else around and I said to the registrar: "call a code blue…"

"A code blue-JAY" I corrected, while running—because that will get the pediatrician and pediatric nurses to respond—and rushed into the ER room: cardiac-2, with Christy Marocick and Dr. Faucettly.

(just as an aside, here several years later I've learned through confidential sources—because I'm now the psych guy—that the pediatrician on call was all fucked up on Percocet himself at the time. No wonder he was just babbling nonsense through the whole code. We all just ignored him assuming it was the stress of a crisis moment. Nope, he was just stoned.)

Dr. Faucettly and I ran the code and we provided "clinically excellent care". Christy Marocick coached us along with her 30 years of Children's Hospital experience. We did the right chest-compression to ventilation ratio and gave all the right doses of ACLS/PALS meds. We analyzed the EKG rhythms correctly and provided the right fucking electricity for each clinically relevant rhythm… I remember Mom screaming out in the hallway and collapsing on the floor when we decided that there would never be an organized rhythm on the monitor again and that we needed to quit trying. The pH was like 6, completely incompatible with life.

We were quietly conversing amongst ourselves, but sure as shit she was listening to our every word.

I say "clinically excellent care" sarcastically because that was the same summer that I won some award thing. I got a plaque that said, "NURSING CLINICAL EXCELLENCE 2007" and I had to eat a rich guy dinner at some golf-club country house with the CEO of the hospital and my picture was hung up in the hallways of all the different buildings of the hospital campus for the whole year. What the fuck does that mean, anyway?

But it was the other nurses that were the real heroes that day. People like Brenda Danzigly, Trisha Brand, and Christy Marocick who provided the emotional support for the family. After we called the code and the family (who was standing around through it all) didn't know what to do, Trisha asked them if they would like to hold their baby one last time. She showed them how to put their arms out and even though they were scared, she helped them to hold and touch and hug and rock their little cold, blue baby one last time.

They were the real 'clinically excellent nurses'. They are the true heroes.

13   Perceptions equal Pain

- Our feelings are created by our thoughts and not the actual events.

- All experiences must be processed through the brain and the powerful limbic-hypothalamic-pituitary-adrenal system.

- The cognitive understanding causes the emotional experience.

- If our understanding of what is happening is accurate, our emotions will be normal.

- If our perception is distorted, our emotional response will be abnormal.

- Depression falls into this category; it is always the result of mental distortions (similar to the static on a TV screen).

- When we learn to bring about this mental fine-tuning, the picture will become clear again.

- Feelings are not facts! Actually, the opposite is true.

- Thoughts create emotions; therefore emotions cannot prove that thoughts are true: we must use logic.

- Mindfulness and Disputation are powerful tools for changing habitual emotional reactions that hijack our ability to think clearly, act skillfully, and live meaningful lives. (These two terms are the meat and potatoes of 'Dialectical Behavioral Therapy' and 'Cognitive Behavioral Therapy, respectively—they have been scientifically proven to be potent techniques).

- They become stronger and more effective as we repeatedly apply them to our lives.

- To increase skills, use these steps each time a negative emotion (irritation, impatience, anxiety, anger) threatens to dominate your awareness:

- Stop

- Bring your awareness to the negative emotion as soon as possible.

- Begin to recognize the early warning signals of the emotional reaction (which are modulated by the limbic-HPA-axis)

- Remind yourself: "I need to pay attention to this—now!"

- Breath

- Become sensitive to the natural softening quality of the breath.

- Use the power of body-mind communication: send a mental message to release and let go

- Relaxing into the exhale, allow the negative emotion to soften.

- Reflect

- Appraise the situation. What is my old pattern here?

- Is my reaction a cognitive distortion?

- What is coming from past experiences/hurts/?

- What resources and options do I have right now?

- Can I change my mind about how I see myself in this situation?

- What is my best insight? What do I want to remember?

- Choose

- Having become more aware of my reaction, settled myself a bit and tapped into my insight, what is possible here?

- Can I shift my old pattern and make a creative choice about speech and actions?

- What is my best choice between this stimulus and response?

14   Is Mental Illness the Real Cause of the Nursing Shortage?

The number one cause of disability worldwide is mental illness. For example, depression is the leading cause of 'lost work days' in the world (World Health Organization, 2010). Vecchio, Scuffham, and Hilton (2009) extrapolate this data to the nursing workforce: mental health problems are the leading cause of disability in nursing and the true cause of the nursing shortage. Even more disturbing is research that shows a direct relation from the number of hours worked in nursing to the rate of mental illness in nurses (Vecchio, Scuffham, & Hilton, 2009).

The real cause for the supposed "nursing shortage" is emotional-dysregulation and a lack of resilience. An effective approach to meeting shortages in the nursing workforce should include strategic attempts to improve mental health capital (Vecchio, Scuffham, & Hilton, 2009). Nursing-career resilience can be affected by three things: 1) emotional reactivity, 2) invalidating circumstances, and 3) learned helplessness. Two techniques taken from Dialectical Behavioral Therapy (DBT) and Cognitive Behavioral Therapy (CBT) can strengthen career resilience. In particular, two techniques: mindfulness and disputation, which are used in the scientifically proven therapies; DBT and CBT, reinforce resilience and mental health capital. We have already touched on mindfulness, but let's dig a little deeper into the problem with nursing and then hit DBT and CBT in a little more detail.

The Problem

The American Association of Colleges of Nursing (Rosseter, 2008), discusses that the shortage of registered nurses (RNs) in the United States could reach between 500,000 and one million by 2025. A large part of the problem is the job stress and dissatisfaction that RNs experience. McVicar (2003) notes there is little doubt that nursing is a stressful profession, and Rosseter (2008) reports that 37% of currently employed RNs are unhappy enough that they are considering changing jobs.

The high rate of chemical abuse and misconduct by nurses is well known, and there is no research available to tell if this is a symptom or a cause of nursing burnout. The number and percentage of nurses with sanctions imposed by state boards of nursing in the United States has risen in the last decade, with male nurses disciplined at a rate twice that of female nurses (Zhong, 2009). In 2005, the American Journal of Nursing suggested that it may simply be post-traumatic stress disorder (PTSD) from on-the-job trauma that is driving nurses from the profession (Schwarz, 2005).

Experienced, knowledgeable bedside nurses are vital to quality health care, but many leave the acute-care setting within five years of starting (Aiken, Clarke, Cheung, Sloane, & Silber, 2003). As experienced nurses leave an institution, institutional memory leaves as well (Hart, 2007, p. 101). This loss of experience creates an unsafe practice environment for new nurses, who then quickly burn out and leave the bedside. This cycle of lost experience propagates the ongoing nursing shortage (Beurhaus, Staiger, & Auerback, 2009).

Vecchio, Scuffham, and Hilton (2009) have shown that the more hours an individual works in the nursing profession, the worse his or her mental health. The worse a nurse's mental health, the fewer hours he or she can work because of disability (Vecchio, Scuffham, & Hilton, 2009). These authors argue quite emphatically that "the findings of this study imply that an effective approach to meeting shortages in the nursing workforce should include strategic attempts to improve mental health capital" (Vecchio, Scuffham, & Hilton, 2009, p. 316).

15 The Problem: Emotional Reactivity and Learned Helplessness

Some nurses stay at the bedside for an entire career, whereas the average nurse leaves acute-care within five years. In fact, the American Journal of Nursing recently reported that the average new-grad RN lasts only 18 months in the acute care setting. This dilemma may demonstrate the different ways people process information about their environment. Even more profound are the individual physiological differences in the temperament of nurses. High sympathetic nervous-system reactivity combined with a difficult environment can lead to debilitating learned helplessness. Learned helpless, which is very similar to clinical depression, may also be a factor leading to nurses leaving the workforce.

Emotional Reactivity

Brandstatter (2001) has done extensive research comparing personality type and emotional responses of people to everyday life situations. Brandstatter (2001) seeks to discover why certain people react differently to various situations. An important part of different reactions are individual personality types, particularly a person's temperament. Temperament is understood as a person's inherited general reactivity (sensitivity) of the nervous and endocrine system (Brandstatter, 2001).

Reactivity is the sympathetic-nervous-system's response to stress. The sympathetic nervous system (SNS) uses the 'fight-or-flight' mechanism when responding to a real (or imagined) threat. Selye's General Adaptation Syndrome has replaced fight-or-flight as a theoretical construct, but the SNS is still to blame. An individual with extremely high reactivity will respond to a minor stressor with an exaggerated increase in heart rate and blood pressure. Cognitive and behavioral interventions are aimed at reducing the sympathetic outflow and mitigate the toxic effects of increased emotional reactivity.

Highly emotionally reactive individuals react poorly to emotional invalidation.

Invalidating environments emphasize

1) controlling emotional expressiveness,

2) oversimplifying the ease of solving problems, and

3) are intolerant of displays of negative affect.

This could be some form of trauma—sexual abuse, etc.

Or, it could be merely extreme reactivity and a cold or unloving home environment.

Think of a little sensitive boy who cries easily and his ruff, tough alcoholic Father smacks him around (whap, whap)

"Big boys don't cry!!!"

An individual's private experiences are responded to erratically and with insensitivity.

The individual learns to mistrust his or her internal states, and instead scans the environment for cues about how to act, think, or feel.

This environment exacerbates the emotional vulnerability and consequent emotion dysregulation

Emotionally invalidating environments create internal pain magnified in intensity by learned helplessness. The degree of emotional reactivity differs for different personality types and can be measured using psychometric tools and even brain-scanning technology. As an example, new research using functional magnetic resonance imaging (fMRI) has shown a difference between how negative or positive visual images are processed by different personality types (Larsen & Buss, 2008).

One well-researched personality spectrum is neuroticism versus extroversion. Canli, et al. (2001), showed neuroticism correlated with increased frontal brain activation to negative-emotion producing images whereas extroversion correlated with increased frontal brain activation to positive-emotion producing images. Nurses with a higher level of neuroticism or high reactivity as a personality trait will have a more pronounced reaction to negativity than peers with lower levels.

Invalidating Circumstances

Linehan (1993) in her book on the treatment of borderline personality disorder (BPD) suggests that there is interplay between two separate components that cause a problem with personality: high emotional reactivity (which is genetically brain-based, i.e. nature) and invalidating circumstances (which are the nurture/environment part of the equation).

Workplace stress in health care is an invalidating environment. Changing the genetics of a nurse's brain is not a feasible workplace intervention. However, changing the environment (i.e. the invalidating circumstance) is a possibility that is within reach of all health care organizations and an individual nurse can learn coping skills to deal with increased emotional reactivity.

Learned Helplessness

Attributional style or explanatory style indicates how people explain to themselves why they experience a particular event as either positive or negative (Seligman, 1992). Learned helplessness is the psychological state in which an individual comes to believe that he or she is powerless to influence the outcome of a situation. Hiroto and Seligman (1975) published seminal research demonstrating that learned helplessness can be produced in humans. The authors define learned helplessness as "inescapable aversive events presented to animals or to men" that results in profound interference with later learning (Hiroto & Seligman, 1975, p. 311).

Learned helplessness looks like clinical depression. Depression is the leading cause of disability as measured by years of lost work, and the 4th leading contributor to the global burden of disease in 2000. By the year 2020, depression is projected to reach 2nd place in the ranking of global burden of disease calculated for all ages, and both sexes (World Health Organization, 2010).

Learned helplessness is the concept describing how individuals come to believe that their actions or behaviors cannot change the outcome or result of life strivings and goals (Schepman & Richmond, 2003). According to Hiroto and Seligman (1975) a recognizable debilitation produced by uncontrollable events is generalized across different motivations and tasks performed. This means that not only does work suffer; but nurses often take traumatic stress home with them (i.e. secondary trauma or vicarious trauma).

But, all these theories about the causes of our bad moods have the tendency to make us victims—because we think the causes result from something beyond our control…….. In contrast, you can learn to change the way you think about things, and you can also change your basic values and beliefs. And when you do you will often experience profound and lasting changes in your mood, outlook, and productivity.

David Burns

Resilience

"Resilience is the process of adapting well in the face of adversity, trauma, tragedy, threats, or even significant sources of stress...It means bouncing back from difficult experiences" (APA, 2010, p. 1).

A growing body of research supports the theoretical model of resilience as an important part of mitigating the effects of stress for bedside nurses in many specialty areas (Gillespie, Chaboyer, Wallis, & Grimbeck, 2007). Hodges, Troyan, and Keeley (2009) researched an exhaustive list of categories and found that building professional resilience "evolved from and seemed to capture the overarching process that led to career longevity among participants" (p. 85). Behavioral techniques that enhance resilience can be used to escape from the learned helplessness fostered by emotional reactivity and invalidating circumstances.

Resilience employs behaviors that facilitate adaptation, decreases dysfunctional behavior, and allows individuals to return to a higher level of functioning after a significant stressor (Gillespie, Chaboyer, Wallis, & Grimbeck, 2007). Hodges, Troyan, and Keeley (2009) show that the natural development cycle of experienced critical-care nurses includes a struggle with professional identity, reconciliation of career choice, values, and everyday practice issues. Nurses must develop self-efficacy and proactive positive adaptability to daily challenges for career longevity in acute-care settings.

Daily hassles and stressors may have an even more pernicious dysregulating effect on well-being than major life events (Montpetit, Bergeman, Deboeck, Tiberio, & Boker, 2010). The average career trajectory of acute-care nurses includes continued self-examination and critically reflective insight applied day-to-day and after profoundly stressful events (Hodges, Troyan, & Keeley, 2009, p. 84).

The American Psychological Association (APA) believes resilience can be taught (APA, 2010) and resilience training has been shown to improve hardiness scores among nurses (Judkins & Ingram, 2002). Resilience is not something that only extraordinary people have, but rather it seems to be one of the defining attributes to being human (Seligman, 2002), (APA, 2010). The transformative power of the individual and growth through adversity, are vital aspects of resilience that allow nurses to deal with ever-changing disruptions in the practice environment (Hodges, Troyan, & Keeley, 2009).

Resilience involves behaviors, thoughts, and actions that are learned and therefore can be developed in anyone (APA, 2010). This dynamic process that "results in adaptation during, or after, significant adversity" (Gillespie, Chaboyer, Wallis, & Grimbeck, 2007, p. 428) may help nurses meet and overcome daily challenges.

Researchers differ on whether resilience is a pre-determined, unchangeable personality trait or a dynamic developmental process (Montpetit, Bergeman, Deboeck, Tiberio, & Boker, 2010). But research has shown that nurses with higher resilience are more emotionally resistant to the detrimental effects of stress (Montpetit, Bergeman, Deboeck, Tiberio, & Boker, 2010).

Resilience provides a much needed paradigm for understanding the nursing shortage and the persistence of nurses who have chosen to stay at the bedside long-term (Hodges, Troyan, & Keeley, 2009). Fostering resilience is the primary theme of strength-based models of therapy within the mental-health field and the positive psychology movement (Atkinson, Martin, & Rankin, 2009). Strength-based models of therapy include Cognitive Behavioral Therapy and Dialectical Behavioral Therapy. Disputation and mindfulness, respectively, are the key techniques used in the two therapies.

Not what you endure, but how you endure, is important.

Seneca

We all know how easy it is to identify with our disappointments and failures. Maltz called it the 'destructive instinct'. Instead of telling ourselves, "I failed to get that job I wanted," we conclude, "I'm a failure." Instead of thinking, "That relationship just didn't work out," we say to ourselves, "Who would want me?"

Psycho-cybernetics, 2000

You can precondition your mind to success. This is a basic principle of positive thinking. You can actually forecast what your future failure or success will be by your present type of thinking

Norman Vincent Peale

Cognitive Behavioral Therapy—Disputation

Thoughts and perceptions of events create feelings and not the event. All experiences are processed through the brain and the powerful limbic-hypothalamic-pituitary-adrenal system (Beck, 1976). The limbic-pituitary-adrenal axis is the tool of the sympathetic nervous system that creates the physiologic response to stress.

The cognitive understanding of an event causes the emotional experience and not the other way around. Often if the understanding of what is happening is accurate, emotions will be normal. Perceptions distorted by incorrect/illogical thinking prompt an emotional response that is abnormal (Burns, 1999). Depression falls into this category; it is always the result of mental distortions (similar to the static on a TV screen) (Burns, 1999).

Disputational resilience is the term that describes how a nurse encounters and interprets stressful experiences (Montpetit, Bergeman, Deboeck, Tiberio, & Boker, 2010, p. 633). Nurses who learn to bring about this mental fine-tuning will notice immediate, profound improvement in stress-mediated mental dysfunction (Burns, 1999). Mental-fine tuning simply means analyzing illogical thinking and arguing against these thinking errors. Beck (1976) calls this technique *disputation* and argues that is one of the most powerfully psychological techniques for healthy living.

Disputation is the foundation of cognitive-behavioral therapy (CBT) and cognitive-behavioral modification can be applied at an organizational level (Boan, 2006). A crucial aspect of becoming professionally resilient is a continual refining of reflective knowledge of self with a sustained commitment to cognitive coherence within the nurses practice environment (Hodges, Troyan, & Keeley, 2009).

Feelings are not facts; the opposite is true. Thoughts create emotions; therefore emotions cannot prove that thoughts are true, instead CBT uses logic to evaluate emotional responses to stressful workplace events (Burn, 1999, Linehan, 1993). Mindfulness and Disputation (Linehan, 1993, Beck, 1976) are powerful tools for changing habitual emotional reactions that hijack the bedside nurse's ability to think clearly, act skillfully, and live meaningful lives.

These techniques become stronger and more useful to the acute care nurse's functioning the more they are practiced. Individuals "construct meaning within social worlds through interaction with others and interpretations they ascribe to situations, a philosophical belief known as symbolic interaction" (Hodges, Troyan, & Keeley, 2009, p. 84). Symbolic interactionism flies in the face of behaviorism in that it assumes nurses can and do think about their actions rather than respond mechanically to stimuli (Hodges, Troyan, & Keeley, 2009).

Not to be able to stop thinking is a dreadful affliction, but we don't realize this because almost everybody is suffering from it, so it is considered normal. This incessant mental noise prevents you from finding that realm of inner stillness....

Tolle

Why is it that so often our first things aren't first?

Covey

Dialectical Behavioral Therapy—Mindfulness

To increase coping skills using mindfulness, some simple steps can be used each time a negative emotion (e.g., irritation, impatience, anxiety, anger) threatens to dominate awareness:

1) Stop—bring cognitive awareness to the negative emotion as soon as possible and begin to recognize the early warning signals of the emotional reaction (modulated by the limbic-HPA-axis)

2) Breathe—become sensitive to the natural softening quality of the breath and use the power of body-mind communication to send a mental message to release and let go.

3) Reflect—appraise the situation. Analyze what the old, dysfunctional cognitive pattern is. Think about whether the reaction is a cognitive distortion. Find out what resources and options are available. Increased nurse longevity in acute-care environments results from these techniques (Hodges, Troyan, & Keeley, 2009).

4) Choose—after becoming more aware of the emotional reaction, settled the sympathetic nervous system response, and tapped into creative insight; try to determine what better outcome is possible in this situation. Analyze the best choice between this stimulus and response (Covey, 1989).

☐　　　The heart of mindfulness is to develop the ability to witness the inner landscape without becoming fully immersed in the stream of spontaneously arising thoughts, emotions, and sensations that constantly flow through us.

☐　　　Experience an emotional arising without it totally dominating the awareness.

☐　　　Witness the inner emotional landscape (without making a judgment)

☐　　　Tone down the sympathetic nervous system

☐　　　Some psychologists compare anxiety and depression, saying that they are the same thing

☐　　　Depression is looking at the past

☐　　　Anxiety is looking at the future

☐　　　Triage is looking at the future (what patient is the most likely to die in the next 2 minutes?)

☐　　　Nurture the ability to settle negative energies.

☐　　　Strong emotional reactions (mediated by the sympathetic nervous system) frequently hijack the ability to think clearly and not act in ways that are consistent with life's best intentions.

Simple technique: 3 slow breaths and clear your mind when you feel the SNS getting revved up.

Try this the next time some asshole cuts you off in traffic and you feel your heart rate go up and you struggle with thoughts of flipping him off, cutting him off with your car, smashing his windshield with a baseball bat, cutting him with a shiv, and hiding his body in the trunk of your car— anyway: mindfulness will work to tone down your sympathetic nervous system's response to stress. Remember,

The sympathetic nervous system, modulated by your limbic system, CAN NOT tell the difference between a real or imagined event!

The number one cause of disability worldwide is mental illness. Depression is the leading cause of 'lost work days' in the world (World Health Organization, 2010) and correlates with the learned-helplessness theoretical construct (Hiroto & Seligman, 1976). This data is extrapolated to the nursing workforce: mental health problems are the leading cause of disability in nursing and the true cause of the nursing shortage (Vecchio, Scuffham, & Hilton, 2009). Even more disturbing is research that has shown that the number of hours worked in nursing is associated with the rate of mental illness in nurses (Vecchio, Scuffham, & Hilton, 2009). Emotional reactivity and invalidating environments can lead to learned helplessness. Resilience research is the study of "intra-individual plasticity in development processes" (Montpetit, Bergeman, Deboeck, Tiberio, & Boker, 2010, p. 631). Acute care nurses can use disputation and mindfulness to develop resilience.

Many situations in healthcare are hopeless and helpless. Many patients are beyond help. Take a pulmonary embolus, for example. Sometimes there is not a single thing we can do but watch and wait, hope and pray.

==

It was slow in the ER and a grizzly, forty-year-old truck driver seemed lonely and wanted to talk. I was bored and so I pulled up a stool next to his bed and sipped on a cup of thick, black coffee.

"Two years ago I fell down dead" the trucker said.

I was suddenly interested and with mock disbelief I said, "really?! what the hell happened?"

"I was having some difficulty catching my breath and suddenly I hit the floor and couldn't get up. I know I died because I was looking down at my brother, who at the time was a state patrol officer; I was looking down at him doing CPR on me".

"I watched him for a little while and I felt a great light come around me and surround me. It was so peaceful. I can't explain the feeling. It was perfectly calm and there was no pain".

There was no pain.

Shit...

The doctor told me later that they worked on me for eight straight hours before they finally got me back. He was a new doc, an intern, and my wife said he didn't leave my bedside for an entire night.

I was in ICU on a vent for a long damn time. I don't remember much about that. I suppose they kept me sedated or something.

But, man it hurt so much.

It was so much damn work to come back. I told my wife that if that ever happened again, I wasn't coming back. That light was too peaceful, too pain free.

Coming back and fighting to live was the hard part.

It was hard work. Lots of pain.

Lots of pain.

It turned out that I had a large P.E. A pulmonary emboli. A huge ass blood clot landed in my lungs and cut off my air. Nothing they can do about it once you got one, you just gotta let your body naturally degrade it, kinda break it down. And you hope you live long enough for that to happen.

Anyway, a year or so after I had my P.E. my brother, the one who was the cop, he fell down dead from the exact same damn thing. Weren't nobody around for him though. Now here I am laying in the ER with the same thing all over again. Doc says I have a "saddle emboli"; it's another huge ass blood clot that is sitting right in the big blood vessel that leaves the heart to go to the lungs. It's sitting right where that big blood vessel branches in two. One branch goes to the right lung and the other to the left. This blood clot is ninety percent blocking both branches and on the cat scan it looks like a saddle. If it moves, or it gets a little bigger, I'm a goner. Doc says I can't move around much, in case I dislodge it, and they got enough blood thinner medicine running in my IV to kill a horse.

Hey buddy?

I want you to know that if I do die again, don't bring me back.

I told my wife I ain't coming back but she'll want everything done, I suppose.

So will the Doc.

But man, I want to stay in the light; where there's no pain… so don't bring me back."

===

19   Greatness

It is important to examine how, without thinking, we use words [cognitions] to empower or demoralize ourselves and others. We can build up and tear down ourselves with the things that we automatically think and say about ourselves. It is also important to understand how the powerful sympathetic nervous system can influence our life. Fear is the root cause of much mental dysfunction and fear is propagated by the nervous system.

Most of us would like to live a peaceful life, but we cannot hope to live happily and peacefully if our lives have been full of violence. Our perceptions and words can be violence to our minds. If our minds have been agitated by emotions like anger or fear we must cultivate peace in our mind, and in our way of life.

The call and need is for greatness. As Stephen Covey says, it is for fulfillment, passionate execution, and significant contribution. Tapping into the higher reaches of human genius and motivation-- requires a new mind-set, a new skill-set, a new tool-set, new automatic cognitions. Words that heal instead of destroy.

20   ER Nursing Permeates Everything

It's weird how a brown furry dog could be both a "purchase" and a "dear friend" after all these years. Standing outside on the back porch, just now; with the train air horns echoing against the wood privacy fence, the stars seemed a little brighter than usual. My dog was sniffing out bunny trails and I enjoyed the cool breeze.

The Union Pacific was on the track through the middle of town, and the Burlington Northern on the North edge of town, both sounding their warning blasts as they approached the multiple intersections with city streets. The horns, like the ER, permeates all of life. Emergency workers are impacted by their work in many ways, sometimes on subconscious levels. Often in ways that influence every aspect of their life.

I have slowly made my peace with the air horns. Working full-time ER nights, those sounds previously felt like warning blasts that life was coming to an end. To me, they signified the prehistoric death-warning like kill-deer birds of western lore.

So many deaths on those tracks. Well, shit. I exaggerate again. Maybe five or six. But they were particularly poignant deaths. Traumatic graves; those steel parallel tracks. The young Mother who took her own life. The Newspaper could not explain what had happened at that calm intersection of track and asphalt. But, we all knew. We had seen her just the week before. Suicidal and psychotic... post-partum. Nothing we could do as she lost her battle with sanity.

The pictures that Brando brought in from another death a few weeks later. A human tongue, juxtaposed from its cranial vault. Just lying on the timbers, on those strong railroad ties, between the parallel steel. "So much bigger out of the mouth" we all said, surprised. The lead engine showed a mere shadow of slime: a spot on the paint where a human head had implanted itself at sixty-mile-an-hour. "Shit, and that! Is that a chest or legs? A torso or toes?"

So much gore, too little time to snap pictures. "Looks like hamburger" I said, my mouth full of a double cheese-burger with no onions, and a diet coke—light on the ice.

"Then one day, while sitting on a plane, headed to God knows where, I had a revelation. I am constantly in the air sitting next to guys who are about my age, and they talk to me as if I am twenty years younger than they are. And they seem twenty years older than I am. They always seem to have sticks up their asses. Where was my stick, I wondered? Where did the stick come from? Was there something inherent in being an adult that I had missed?

Lewis Black

So, I promised myself that I would write about my first shift back in the ER after an extended hiatus. By "extended", I mean like 2 months. The first of those last two months, I spent talking about mental illness. I was lecturing six days a week about insanity, and it had started to affect my sanity. The last month of those two months, I spent sitting and thinking.

I sat and tinkered on my old, vintage Harley and I read, read, read. Drank a little beer, smoked some Camel Menthols, listened to 103.1 (hard/emo rock), and thought about shit.

I mostly read the wiring schematics for the Ironhead Sportster, but I also read "Zen and the Art of Stand-up Comedy", "Zen and the Art of Motorcycle Maintenance", "The Stranger", "How the Mind Works", "The Power of Now", and "A Twist of the Wrist".

I love to read. I did a little writing as well. Got myself a "comedy journal" in which I write (as brutally honest as possible) everything that I find hysterical in my life. I hope to do a 3-minute open mic night in Indianapolis this week. The topic? FEAR--and how it has ruled and ruined my life to date. Pretty funny shit... haha...

So, anyway...

My 12-hour shift in the ER today was laid-back and thus: fucking great! I worked with cool, beautiful people all day. By "beautiful", I mean 'nice and genuine'. We saw some shit, but not much, really. Every single patient that comes to the ER is having a 'bad day'. Every single ER visit is a very big deal to the person visiting and it's important for us as nurses to empathize with this perceived catastrophe. However, if I had to categorize the last nine patients that I had; room-by-room, it would be:

Exam room 1) bullshit,

Exam room 2) bullshit,

Exam room 3) bullshit,

Exam room 4) bullshit--but cute and chubby!

Exam room 5) bullshit,

Exam room 6) bullshit,

Exam room 7) bullshit--but nice to talk to,

Exam room 8) bullshit,

Exam room 9) bull... OH SHIT!!!!@#$!!!

Awesome. That's why we love the ER.

==

"We must look at the lens through which we see the world, as well as at the world we see, and that the lens itself shapes how we interpret the world"

Covey

How can I reinvent myself? I often ask myself this in times of stress, crisis, boredom, or mandatory staff meetings in which everyone else is pretending to give a shit about the nothing that is being said.

I first began thinking of this question some years ago when I received a phone call from the doctor's office. I was 30 years old, and the polite nurse just wanted to read me the results of a chest x-ray done because of increasing shortness of breath.

The dyspnea had become severe enough that it was impacting my 12-hour shifts in the ER, and I would often spend my days at home lying on the couch without the energy to decrease the elevation of my feet from the same horizontal plane as my head.

I could not even get up.

I was sleeping at night with 3 or 4 pillows to prop myself up and a fan blowing full force at my face. My poor wife, who is always too cold, would cover herself with about five blankets wrapped up like a mummy.

Often the mere cotton sheet on my side of the bed felt suffocating to me.

The nurse, said, "Josh the doctor just wanted you to know the results of your x-ray… It says, hmmm… let's see… Oh, here it is: Emphysema".

"You have emphysema".

She paused, I paused…

And then I said: I don't know why, but I said it enthusiastically "Well, thank you for calling!"

The cheerful response: "Have a nice day".

I am an experienced critical care nurse, but at that exact point in time I could not remember what emphysema was. Like any good nurse, I went to the first reputable site on the internet that I could find: Wikipedia. I read a bit about emphysema and decided to stop with the sentence: "Often individuals who are unfortunate enough to contract this disease have a very short life expectancy, often 0–3 years at most".

Shit.

I still had a lot of living to do and reality had fucked me again.

I have learned that if one advances confidently in the direction of his dreams, and endeavors to live the life he has imagined, he will meet with a success unexpected in common hours.

Henry David Thoreau

I cut out and pasted on the bathroom mirror this Thoreau quote:

I know of no more encouraging fact than the unquestionable ability of man to elevate his life by conscious endeavor.

I also found this Thoreau quote:

How many a man has dated a new era in his life from the reading of a book

And so I started reading. I learned that from Drew Carey, he says that self-help books saved his life. He attempted suicide several times in high-school. The first book was suggested by an ER doc friend who had used it to quit smoking: Napolean Hill's "Think and Grow Rich".

Then I had to do the difficult work of

1) Trying to figure out what my dreams actually were
2) How to go about accomplishing them in a few short years
3) What things were getting in my way?

Question number three was the hardest. I worked closely with a pulmonologist. I did allergy testing, pulmonary function testing, tired a plethora of medications, and she decided that it was not emphysema (an always fatal lung disease--thanks Wikipedia). I had two separate problems: vocal cord dysfunction (VCD), and allergy/exercise induced asthma.

A wonderful doctor is worth more than… shit, a nice metaphor escapes me. Worth more than something really, really, valuable. I now spend well over a hundred dollars a month on a handful of inhalers and pills and feel wonderful (for the most part—but don't ask me to mow the lawn—grass is nearly fatal for me). Interestingly, VCD is found in 1) Nurses 2) Doctor's wives 3) Borderline Personality Disorder (high emotional reactivity). I am the first and third. Shit, was I just somatizing? I am not a physician's wife because things just never worked out that way.

What I realized is that it is my own mind and

FEAR that has always prevented me from maximizing my potential.

The band Hurt in the song "Falls Apart" spoke of this:

"I messed up again when I tried
You spend all your money and die.
And, oh! By the way,
With all you did nothing has changed
So lie like a waste by the side

As everything just falls apart
'Cause everything just fell apart for me"

I struggled with suicidal ideation (along with the other 20 million Americans who struggle each year with this existential crises). Asthmatics have a 30% rate of suicidality. Funny how the inability to breathe, which is at the absolute bottom of Maslow's hierarchy of needs, prevents you from moving up the pyramid to extreme self-actualization.

I started running.

At first it was a brisk walk around the block. I found that the racing thoughts of hypomanic dyspnea only worsened. Back to the drawing board. I read more books. I slowly pedaled on a recumbent exercise bike downstairs and read David Burns "Feeling Good: The New Mood Therapy".

I learned the phrase "automatic cognitive distortion".

It showed me how my thinking was essentially flawed. This helped a lot, but my mind still spiraled down out of control, at times catastrophically, with seemingly nothing I could do. I had fearful thoughts. Air became a monster to fight against. Air carried pollens and demons that were somehow attacking my lung tissue.

Bronchitis frequently occurred. The definition of "Chronic Bronchitis" is coughing up ½ a cup of phlegm every day for 3 months out of the year. Disgusting. My poor wife.

Dr. McFan, an ER coworker and good friend, would always make me take a shot of Rocephin when I would have these flare ups at work. I refused to let any of my coworkers see my bare ass and so I would shoot up in the leg in the men's restroom. One time (after watching this on an episode of 'ER'—remember the narcotic-addict physician) I dropped the empty syringe out from under the bathroom stall and let it roll on the floor a bit. I came out of the stall, picked up the syringe and quietly left. It was hysterical. I had to pee in a cup later that night.

I religiously took my meds and discovered "Mindfulness Running"

http://www.mindfulness.com/category/mindful-running/

This revolutionized my life physically.

As a single footstep will not make a path on the earth, so a single thought will not make a pathway in the mind. To make a deep physical path, we walk again and again. To make a deep mental path, we must think over and over the kind of thoughts we wish to dominate our lives.

Henry David Thoreau

I have spent the last 3 years running a couple of times a week. Slowly increasing my distance, and slowly learning to clear my mind of racing thoughts. It was not fast progress, but I realize now that I have changed for the better. I have slowly tried to change the negative thoughts to positive ones—just like all the self-help books suggest.

I ran four miles across a beautiful bridge over the Tennessee River in Chattanooga. Homeless guys drinking Olde English 800 laughed at "that white-boy running so slow". I smiled and waved. They smiled and waved back.

Sometimes; serene, mindfulness is so valuable. Sometimes, (like yesterday) I put in another self-help audiobook to gain valuable advice about creating a life worth living.

Spirituality can be important as well.

"I've been to both knees
Raise my hands up to the sky, forgive me.
Is something out there far beyond the clouds
I'm asking help me
Help me to see the world
Through baby eyes and hold me closely"

"I want to shine like that
I want to smile so big my daughter jumps into my lap
And I wanna tell her daddy's fine and always plans to be"
Blue October "Blue Skies"

Yesterday it was Timothy Ferris' "The 4-Hour Workweek" a sort of trite, modern-day Walden. I ran 10-miles. Not fast, about 11 minute miles, but still… I believe, mindfulness has helped me tremendously…

References

Aiken, L., Clarke, S., Cheung, R., Sloane, D., & Silber, J.H. (2003). Educational levels of hospital nurses and surgical patient mortality. Journal of the American Medical Association, 290, 1617-1623.

American Psychological Association (APA). (2010). The road to resilience. APA Help Center. Retrieved from: http://www.apa.org/helpcenter/road-resilience.aspx

Atkinson, P.A., Martin, C.R., & Rankin, J. (2009). Resilience revisisted. Journal of Psychiatric and Mental Health Nursing, 16, 137-145.

Beck, A.T. (1976). Cognitive therapy and the emotional disorders. New York: Penguin Books.

Boan, D.M. (2006). Cognitve-behavior modification and organizational culture. Consulting Psychology Journal: Practice and Research, 58(1), 51-61.

Brandstatter, H., & Eliasz, A. (Eds.). (2001). Persons, situations, and emotions: An ecological approach. Oxford: Oxford University Press.

Buerhaus, P., Staiger, D., & Auerbach, D. (2009). The future of the nursing workforce in the United States: Data, trends, and implications. Boston: Jones & Bartlett.

Burns, D.D. (1999). Feeling good: The new mood therapy. New York: Harper.

Canli, T., Zuo, Z., Desmond, J.E., Kang, E., Gross, J., & Gabrieli, J.D.E. (2001). An fMRI study of personality influences on brain reactivity to emotional stimuli. Behavioral Neuroscience 115(1), 33-42. Retrieved from http://www.apa.org/journals/features/bne115133.pdf

Centers for Disease Control and Prevention --> QuickStats (2007) 56(42);1115, in turn citing: Kung HC, Hoyert DL, Xu JQ, Murphy, SL. E-stat deaths: preliminary data for 2005 health E-stats. Hyattsville, MD: US Department of Health and Human Services, CDC; 2007. Available at http://www.cdc.gov/nchs/data/hestat/prelimdeaths05/preliminarydea

Covey, S.R. (1989). The 7 habits of highly effective people: Powerful lessons in personal change. New York: Simon and Schuster.

Csikszentmihalyi, M. (1991). Flow: The psychology of optimal experience. New York: Harper.

Dyck, L.R., Caron, A., & Aron, D. (2006). Working on the positive emotional attractor through training in health care. Journal of Management Development, 25(7), 671-688.

Gillespie, B.M., Chaboyer, W., Wallis, M., & Grimbeck, P. (2007). Resilience in the operating room: developing and testing of a resilience model. Journal of Advanced Nursing, 59(4), 427-438.

Greengrass, M. (2004). 100 years of B.F. Skinner. Monitor on psychology, 35, 80.

Hart, K. (2007). The aging workforce: Implications for health care organizations. Nursing Economics, 25(2), 101-102.

Hiroto, D.S. & Seligman, M.E.P. (1975). Generality of learned helplessness in man. Journal of Personality and Social Psychology, 31(2), 311-327.

Hodges, H.F., Troyan, P.J., & Keeley, A.C. (2009). Career persistence in baccalaureate-prepared acute care nurses. Journal of Nursing Scholarship, 42(1), 83-91.

Judkins, S. and Ingram, A. (2002). Decreasing stress among nurse managers: A long-term solution. Journal of Continuing Education in Nursing, 33(6), 259-264.

Larsen, R.J., & Buss, D.M. (2008). Pesonality psychology: Domains of knowledge about human nature (3rd ed.).New York: McGraw-Hill.

Linehan, M. (1993). Skills training manual for treating borderline personality disorder. New York: Guilford Press.

McVicar, A. (2003). Workplace stress in nursing. Journal of Advanced Nursing, 44(6), 633-642.

Montpetit, M.A., Bergeman, C.S., Deboeck, P.R., Tiberio, S.S., & Boker, S.M. (2010). Resilience-as-process: Negative affect, stress, and coupled dynamical systems. Psychology and Aging, 25(3), 631-640.

Rosseter, R.J. (2008). Nursing shortage fact sheet. The American Association of Colleges of Nursing. September 29, 2008.

Seligman, M.E.P. (1992). Learned optimism: How to change your mind and your life. New York: Pocket Books.

Seligman, M. (2002). Authentic happiness: Using the new positive psychology to realize your potential to realize your potential for lasting fulfillment. New York: Free Press.

Seligman, M.E.P., & Csikszentmihalyi, M. (Eds.). (2000). Positive psychology. [Special issue] American Psychologist, 55(1).

Seligman, M.E.P., Steen, T.A., Park, N., & Peterson, C. (2005). Positive psychology progress: Empirical validation of interventions. American Psychologist, 60(5), 410-421.

Schepman, S.B. & Richmond, L. (2003). Employee expectations and motivation: An application from the "learned helplessness" paradigm. The Journal of American Academy of Business,3, 405-408.

Schwarz, T. (2005). PTSD in nurses: On-the-job trauma may be driving nurses from the profession. American Journal of Nursing,105, 13.

Skinner, B.F. (2005). A brief survey of operant behavior. Retrieved May 29, 2009 from http://www.bfskinner.org/brief_survey.html

Skinner, B.F. (2005). Science and human behavior. Retrieved May 29, 2009 http://www.bfskinner.macwebsitebuilder.com/f/Science_and_Human_Behavior.pdf

Skinner, B.F. (1958). Teaching machines. Science, 128, 969-977.

Vargas, J.S. (2008). A brief biography of B.F. Skinner. Retrieved May 29, 2009 http://www.bfskinner.org/brief_biography.html

Vecchio, N., Scuffham, P.A., & Hilton, M.F. (2009). Mental health and hours worked among nurses. American Journal of Labour Economics, 12(3), 299-320.

Viney, W., & King, D.B. (2003). A history of psychology: Ideas and context (3rd ed.). New York: Pearson.

Wang, P.S., Beck, A.L., Berglund, P., McKenas, D.K., Pronk, N.P., Simon, G.E., et al. (2004). Effects of major depression on moment-in-time work performance. The American Journal of Psychiatry, 161(10), 1885-1891.

World Health Organization (WHO). (2010). Depression. Retrieved from: http://www.who.int/mental_health/management/depression/definition/en/

Zhong, E.H., Kenward, K., Sheets, V.R., Doherty, M.E., Gross, L. (2009). Probation and recidivism: Remediation among disciplined nurses in six states. American Journal of Nursing, 109, 48-58.